Inspire Science

Ecosystems

McGraw Hill Education

FRONT COVER: (t)Carlos Ángel Vázquez Tena/EyeEm/Getty Images, (b)Dave Massey/
Shutterstock; BACK COVER: Carlos Ángel Vázquez Tena/EyeEm/Getty Images

Mheducation.com/prek-12

STEM McGraw-Hill is committed to providing
instructional materials in Science, Technology,
Engineering, and Mathematics (STEM) that give all
students a solid foundation, one that prepares them
for college and careers in the 21st century.

Send all inquiries to:
McGraw-Hill Education
8787 Orion Place
Columbus, OH 43240

ISBN: 978-0-07-699676-6
MHID: 0-07-699676-X

Printed in the United States of America.

5 6 7 8 9 10 11 LWI 26 25 24 23 22 21 20

Table of Contents
Unit 2: Ecosystems

Matter in Ecosystems

Energy in Ecosystems

Matter in Ecosystems

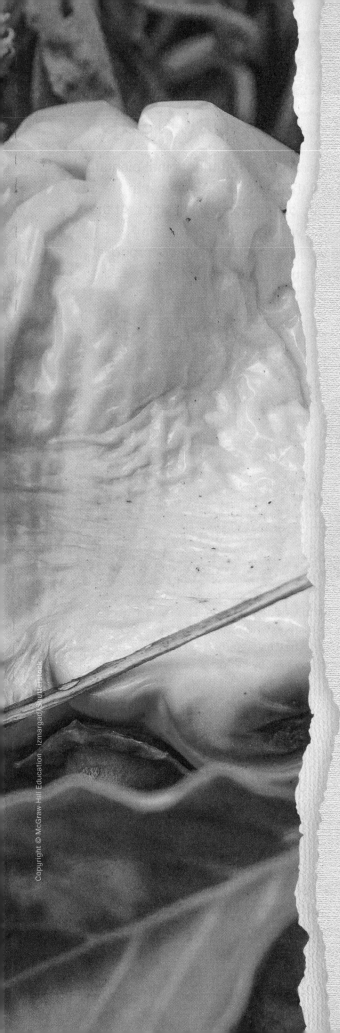

Copyright © McGraw-Hill Education (izmargad)/Shutterstock

ENCOUNTER
THE PHENOMENON

What causes the remains of this plant to change over time?

Rotting Vegetables

⟐ GO ONLINE
Check out *Rotting Vegetables* to see the phenomenon in action.

💬 Talk About It

Look at the photo and watch the video of the vegetables decomposing. What questions do you have about the phenomenon? Talk about them with a partner.

Did You Know?

About thirty percent of what we throw away comes from food scraps and yard waste.

Design a Compost Heap

How can composting provide the nutrients that plants need to grow? You will think like a horticulturist to research different types of composting activities. At the end of the module, you will develop a plan for a compost heap. Use your research and write a proposal for a composting program for your school or home.

Lesson 1
Plant Survival

Lesson 2
Interactions of Living Things

Lesson 3
Role of Decomposers

Horticulturists use their knowledge of plants, soil conditions, irrigation systems and tools to grow a variety of plants. They also study ways to improve plant production by keeping plants healthy.

STEM Module Project

Plan and Complete the Science Challenge Use what you learn about matter in ecosystems to complete your design!

Plant Growth

Four friends were visiting a Botanic Garden. They wondered how the plants at the garden grow. They each had a different idea:

Soto: *Plants make their food from things they get from the soil. They use the food they make to grow.*

Audrey: *Plants make their food mainly from air and water. They use the food they make to grow.*

Belkis: *Plants make their food from soil, air, and water. They use sunlight to grow.*

Carl: *Plants make food mainly from air and water. They don't use the food they make to grow.*

Who do you think has the best idea? _____

Explain your thinking.

_____

_____

_____

You will revisit the Page Keeley Science Probe later in the lesson.

Plant Survival

How does this tree get what it needs to survive?

GO ONLINE

Check out *Giant Sequoia Growth* to see the phenomenon in action.

Look at the photo and watch the video of a giant sequoia tree. What did you observe? What questions do you have about the phenomenon? Record or illustrate your thoughts below.

Did You Know?

The trunk of a giant sequoia can weigh about 1,300 tons, which is equal to the weight of around 650 cars.

INQUIRY ACTIVITY

Simulation

Virtual Plant

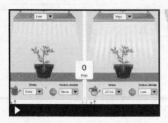

🕑 GO ONLINE

Think about what the giant sequoia needs to grow. Conduct the simulation to investigate the growth of a virtual plant.

Make a Prediction How will the amount of water, sunlight, and air affect plant growth?

_____

_____

_____

Carry Out an Investigation

1. On one plant, set the light and air to high, and water to 40 mL. On the other plant, set the light and air to high but set the water to 0 mL and then 20 mL. Do you notice any patterns? Explain.

_____

_____

_____

2. Now investigate the effect of air. Change the amount of carbon dioxide that one plant gets. What do you observe?

_____

_____

_____

3. What happens when you set all plants to the same setting, but vary the amount of light?

_____

_____

_____

4. **MATH Connection** Parker investigated how the amount of sunlight affects plant growth. Using his data below, calculate the average growth of each plant. Assume that each plant was provided 20 mL of water per day.

	Amount of Sunlight Per Day	Height in Week 1	Height in Week 2	Height in Week 3	Average
Plant A	4 hours	1 cm	3 cm	6 cm	
Plant B	8 hours	1.5 cm	4 cm	8 cm	
Plant C	16 hours	1 cm	2 cm	3 cm	

Communicate Information

5. Which conditions favored the most growth?

_____

_____

6. Which plant had the least growth? What can you infer from those results?

_____

_____

_____

💬 Talk About It

Compare your results with a partner's results. How do the results relate to what you observed about the needs of plants? What other factors could affect plant growth?

> ▶ **GO ONLINE** Explore *Mass of a Tree* to learn about how giant sequoias get to be so big.

Plant Structures

VOCABULARY

Look for these words as you read:

energy

phloem

stomata

transpiration

xylem

The giant sequoia has the same needs as other plants: water, air, sunlight, space, and nutrients. Nutrients are substances that a living thing needs to stay healthy. Plants need energy to meet these needs. **Energy** is the ability to perform work or change something. Plants use structures such as leaves, stems, and roots to obtain energy.

Plants need carbon dioxide from the air to make their food. Tiny openings on the underside of most leaves allow air to enter. These openings—called **stomata**—can close and prevent water from escaping.

Other plant structures obtain the other materials that the plant needs. Water is absorbed by a plant's roots. It travels up the center of the stem through specialized tissues called **xylem**.

GO ONLINE Watch the video *Plant Structures* to see more examples of the different types of plants and their parts.

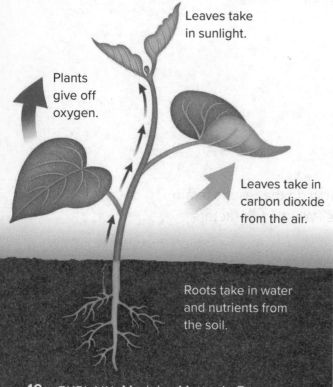

Leaves take in sunlight.

Plants give off oxygen.

Leaves take in carbon dioxide from the air.

Roots take in water and nutrients from the soil.

Explain how the diagram **provides evidence** of the major roles of the different plant parts and the **flow of energy**, water, and air.

_____

_____

_____

_____

_____

_____

Plant Needs

Plants have basic needs to live and grow. You now know that plants use their structures to obtain what they need. Plants need enough space where they grow for their roots to spread out and absorb water and nutrients from the soil.

Water and carbon dioxide combine in the presence of light energy to produce sugar and oxygen. Energy from the sunlight is now contained in the sugars. The sugars are what the plant uses for food. The food is then available to the plant for growth, storage, and other life processes.

Phloem are tissues that transport sugars to all parts of the plant. It drives the movement of materials throughout a plant. **Transpiration** is the evaporation of water from a plant's leaves. As water evaporates from the leaves, more water is carried from the bottom of the plant to the top. Water moves into the leaf, replacing the water that has evaporated.

Some plants need more water than others. Cacti can survive in deserts with little rain, while the plants in a rain forest live in a very wet area.

1. Why would it be a disadvantage if plants grow too close together?

   _____

   _____

   _____

2. Some woody vines can grow on rainforest trees and climb high into the tree canopy. Why would this be an advantage?

   _____

   _____

   _____

INQUIRY ACTIVITY

Hands On

Plant Investigation

Plant parts obtain the resources that the plant needs. You are going to investigate how soil affects plant growth.

Make a Prediction How will using soil, gravel, or sand affect how well a plant can grow?

_____

_____

_____

Carry Out an Investigation

BE CAREFUL Wear safety goggles and gloves.

1. Plan your investigation below. Include any safety statements. Remember to have one cup with cotton balls to serve as the control variable with no soil.

_____

_____

_____

_____

_____

_____

_____

Copyright © McGraw-Hill Education (4 8 10)Jacques Cornell/McGraw-Hill Education, (7)McGraw-Hill Education, (9)Janette Beckman/McGraw-Hill Education, (others)Ken Cavanagh/McGraw-Hill Education

Materials

 safety goggles

 4 plastic cups

 soil

 gravel

 sand

 cotton balls

 seeds

 water

 graduated beaker

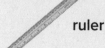 ruler

2. Record the growth of each plant in the table.

	Soil	Gravel	Sand	No Soil (Cotton)
Starting Height (cm)				
Growth After 8 Days				
Growth After 11 Days				
Growth After 14 Days				

Communicate Information

3. MATH Connection   Make a line graph of the data, with a different line color for each type of soil. Label the vertical axis 'Growth' and the horizontal axis 'Days.'

Talk About It

Compare your results with a partner. Did your results support your prediction? How can you improve your investigation?

REVISIT Revisit the Page Keeley Science Probe on page 5.

PAGE KEELEY
SCIENCE
PROBES

What Does an Agricultural and Food Science Technician Do?

Agricultural and food science technicians

work with food and how it is produced. Some technicians study crops. They monitor the quality of the soil and check for pests that can destroy the plants. They also use technology to find ways to package and distribute these foods to people.

Other agricultural and food science technicians work with livestock. They might work to prevent and cure illnesses that are common in livestock. They also research ways to improve how meat from livestock is packaged and transported. Through the use of technology, the technicians learn new ways to maintain the quality of meat so that it is safe to eat.

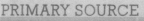
PRIMARY SOURCE

It's Your Turn

Think like an agricultural and food science technician. What information and advice could you give a horticulturist when working together?

INQUIRY ACTIVITY

Soil-less Gardens

Agricultural and food science technicians might study how to grow food crops without soil. Research ways of growing plants without soil by reading the Investigator article *Soil-less Gardens*, going online to teacher-approved websites, or by finding books on hydroponics at your local library.

WRITING **Connection** Write a persuasive argument for why plants should be grown with or without soil by presenting evidence to support your point of view.

_____

_____

_____

_____

_____

_____

_____

_____

_____

_____

_____

_____

_____

Review

EXPLAIN
THE PHENOMENON

How does this tree get what it needs to survive?

Summarize It

Support the argument that the giant sequoia gets what it needs to grow and survive mostly from air and water.

_____

_____

_____

_____

_____

_____

_____

REVISIT

PAGE KEELEY SCIENCE PROBES

Revisit the Page Keeley Science Probe on page 5. Has your thinking changed? If so, explain how it has changed.

Three-Dimensional Thinking

1. Which is found inside the stem of a plant?

    A. epidermis

    B. root hairs

    C. xylem

    D. leaves

2. Explain the process of transpiration.

_____

_____

_____

_____

_____

_____

_____

_____

_____

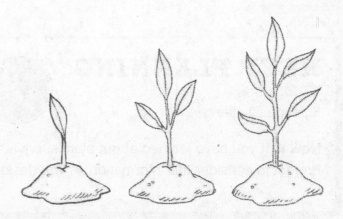

Extend It

Think about what it would be like to live in a world without soil to grow plants. Use what you have learned in this lesson about plant survival to communicate a plan to your community for growing plants in a world without soil. Provide details about where the plants could grow and what they would need to survive. Use the drawing box to sketch your plan, if you need to.

KEEP PLANNING

STEM Module Project
Science Challenge

Now that you have learned about plant survival, go to your Module Project to consider this information as you design a compost heap.

What Do Animals Eat?

Three friends were talking about animals and the food they eat.
They each had different ideas. This is what they said:

Molly: *I think animals interact with matter in their environment to get food
from plants.*

Patty: *I think animals interact with matter in their environment to get food
from other animals.*

Devon: *I think animals get food from plants, animals, or both, depending on
how they interact with matter in their environment.*

Who do you agree with most? _____

Explain why you agree.

_____

_____

_____

_____

_____

You will revisit the Page Keeley Science Probe later in the lesson.

Interactions of Living Things

What is the relationship between these ants and the leaves?

🔘 GO ONLINE

Check out *Leafcutter Ants* to see
the phenomenon in action.

Look at the photo and watch the video of the leafcutter ants. What did
you observe? What questions do you have about the phenomenon?
Record or illustrate your thoughts below.

Did You Know?

Leafcutter ants use the leaves they cut to grow
their own food underground. Each piece of
leaf can be 50 times the body weight
of each ant!

INQUIRY ACTIVITY

Data Analysis

Foxes and Rabbits

Copyright © McGraw-Hill Education (1)Janette Beckman/McGraw-Hill Education, (2 4)Jacques Cornell/McGraw-Hill Education, (3)McGraw-Hill Education

Think about how the leafcutter ants interacted with the leaves. Living things in ecosystems interact with one another.

Make a Prediction What would happen to the population of rabbits if the fox population increased in a forest ecosystem?

_____

_____

_____

Materials

meterstick

masking tape

8 7.5-cm cardboard squares

100 2.5-cm construction paper squares

Carry Out an Investigation

1. Use the meterstick and tape to mark off a 50-cm square. This square represents a forest. Distribute 20 of the small squares within the forest. These squares represent rabbits.

2. The larger squares represent foxes. The fox must touch at least three rabbit squares to live. If one fox touches three or more rabbits, you will toss one more fox in for the next trial.

3. **Record Data** Toss one fox into the forest. Remove any rabbits that the fox touches. Record the results in the data table on the next page.

4. **MATH Connection** Multiply the number of rabbits remaining by 0.3. You may need to round the number of rabbits to the nearest whole number. Record this number under 'Number of New Rabbits in Next Trial.' Place these new rabbits in the forest and complete the next trial.

5. **Record Data** An adult fox survives if it catches at least three rabbit squares. Record the number under "Surviving Adult Foxes" for each trial.

6. **Record Data** If a fox survives it also produces one offspring. Record this number under "Fox Offspring." Add the number of surviving adult foxes and the number of offspring. Record this number under "Beginning Number of Foxes" for the next trial.

- If the entire rabbit population is removed by the fox, add three new rabbits to the forest to represent new rabbits moving into the area. If all your foxes starve, then add a fox to represent a new fox moving into the area.

- In each additional trial, throw each fox square once. This includes any surviving foxes from previous trials and any additional foxes if enough rabbits were caught, per Step 2.

Record Data

Trial	Beginning Number of Rabbits	Beginning Number of Foxes	Number of Rabbits Caught	Number of New Rabbits in Next Trial	Surviving Adult Foxes	Fox Offspring
1						
2						
3						
4						
5						
6						

INQUIRY ACTIVITY

Communicate Information

7. How did the population of foxes change as the population of rabbits increased?

_____

_____

_____

_____

_____

_____

8. What happened to the population of rabbits as the population of foxes increased? Did your results support your prediction?

_____

_____

_____

_____

_____

_____

_____

9. What would happen if the plant population in the forest decreased?

_____

_____

_____

_____

_____

MAKE YOUR CLAIM

What do living things rely on for survival?

Make your claim. Use your investigation.

CLAIM

Living things rely on _____.

Cite evidence from the lesson.

EVIDENCE

The investigation showed that _____.

Discuss your reasoning as a class. Tell about your discussion.

REASONING

The evidence supports the claim because _____.

You will revisit the claim to add more evidence later in this lesson.

Ecosystems

Look for these
words as you read:

abiotic factor

biotic factor

habitat

invasive species

predator

prey

You investigated how the foxes and rabbits in a forest ecosystem interact and affect each other's populations. An ecosystem is made up of all living and nonliving things in an environment. All of the living things in an environment are called **biotic factors**. Plants, animals, fungi, and bacteria are biotic factors. **Abiotic factors** are the nonliving things in the environment. Air, water, soil, rocks, and light are abiotic factors. Both biotic and abiotic factors interact with one another.

 Use information from models you see in the lesson to provide examples of biotic and abiotic factors in **ecosystems** and ways that they interact with each other.

Biotic Factors	Abiotic Factors

Habitats

Ecosystems can be small, like a single log or
a pond, or very large, like a forest or a desert.
Each organism must have its own space.
The place in an ecosystem where an organism
lives is its **habitat**. Habitats vary depending on
the type of ecosystem, such as freshwater habitats like ponds and rivers.
Each living thing has its own niche, or special role that an organism
plays in the ecosystem. For example, an earthworm's niche in
a forest ecosystem is to break down plant matter in the soil.

GO ONLINE Watch the video
The Movement of Matter in an Ecosystem.
Record examples of biotic and abiotic
factors from the video in your table.

Circle the biotic factors in this pond ecosystem.

Circle the abiotic factors in this river ecosystem.

Inspect

Read the passage *Invasive Species.* Underline text evidence that tells how invasive species affect an ecosystem.

Find Evidence

Reread the passage. Highlight the words that helped you determine the meaning of the words *invasive species.*

Notes

_____

_____

_____

_____

_____

_____

_____

_____

_____

_____

_____

_____

Invasive Species

Humans may move an organism from its natural ecosystem to another. If the organism lives and reproduces in the new ecosystem, it can cause harm to that area. An organism that is introduced to a new ecosystem and causes harm is an **invasive species**. Invasive species can harm the environment, the economy, and even human health. Species that grow and reproduce without other animals that hunt it are likely to spread quickly and become invasive.

Sometimes, an invasive species is accidentally introduced to an environment. Other times, it is introduced on purpose. The cane toad was introduced to Australia in the 1930s. A type of beetle in Australia was eating sugar cane crops. Cane toads are known to eat large amounts of beetles. So, farmers moved the cane toad from their natural habitat in South America to the sugar cane fields in Australia. These toads have a toxic skin and have no animal that hunts it for food in Australia. The population is now in the millions! These toads are both poisoning and competing with native species.

Copyright © McGraw-Hill Education Matteo Colombo/Flickr RF/Getty Images

Invasive species are primarily spread by human activities, often unintentionally. People, and the goods they use, travel around the world very quickly, and they often carry uninvited species with them.

Think about the relationship of the fox and rabbit population you explored earlier. What would happen if an invasive species of plant was introduced into the ecosystem that competed with the rabbits' food source?

_____

_____

_____

_____

_____

_____

Make Connections

 Talk About It

What can we do to help limit the spread of invasive species?

Notes

_____

_____

_____

_____

_____

_____

_____

_____

_____

_____

_____

_____

_____

Predator and Prey

In an ecosystem, there are many relationships and interactions of biotic and abiotic factors that overlap. All living things need water, air, and space. Plants are producers and make their own food. Animals are consumers and need food from other sources. Herbivores eat plants, carnivores eat other animals, and omnivores eat both plants and animals.

Organisms that hunt for their food are **predators**. The organisms they hunt are **prey**. Predators are important in ecosystems because they help control the size of prey populations. When the populations of prey are controlled, producers and other resources are less likely to run out.

COLLECT EVIDENCE

Add evidence to your claim on page 25 about how living things interact with other matter in their environment. Use this evidence to provide further reasoning for your claim.

REVISIT Revisit the Page Keeley Science Probe on page 19.

PAGE KEELEY
SCIENCE
PROBES

What Does an Entomologist Do?

Dr. Edward O. Wilson, scientist and teacher, has spent his life peeking into the nests of ants. He's curious about the job of each ant in the colony. He wants to uncover the secrets of ant colonies' success. Dr. Wilson is an **entomologist**— a scientist who studies insects.

Ants live almost everywhere, from tropical climates to beyond the Arctic Circle, from dry deserts to shady rain forests, from city sidewalks to wild woodlands, and from deep in the ground to the tops of the tallest trees. They live in colonies. An ant colony can have as many as 20 million members.

For fifty years, Dr. Wilson has traveled around the world looking for new kinds of ants. Sometimes he brings entire colonies back to his laboratory to observe them more closely. He wants to learn about each ant's job within its colony. He wants to know how each ant's job contributes to the future survival of its species.

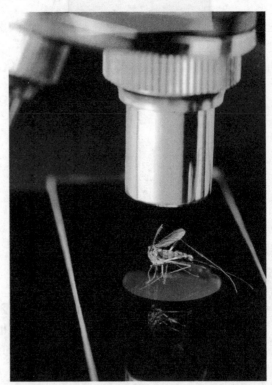

This student is training to be an entomologist. They are collecting samples of insects in an ecosystem to study.

Dr. Wilson's discoveries help us understand why many animal species develop social organization. In a social organization, each member of the group has a specific job. Each job is important to the entire species' success.

Whenever possible, Dr. Wilson returns to the place where he first watched ants. He notes the changes in ant species that have occurred over the past sixty years. And today, he still relies on the observations and collections of specimens that he made when he was a young boy.

It's Your Turn

How can an entomologist and a horticulturist work together? Research more information about the roles of these careers. Write a report on how the work of an entomologist could help the work of a horticulturist.

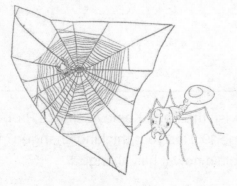

Review

EXPLAIN
THE PHENOMENON

What is the relationship between these ants and the leaves?

Summarize It

Explain how living things interact with one another in their ecosystems, such as the leafcutter ants and plants.

_____

_____

_____

_____

_____

_____

_____

_____

REVISIT Revisit the Page Keeley Science Probe on page 19. Has your thinking changed? If so, explain how it has changed.

PAGE KEELEY
SCIENCE
PROBES

Three—Dimensional Thinking

1. An organism's role in an ecosystem is its _____.

    A. habitat

    B. niche

    C. producer

    D. prey

2. Think about a marine ecosystem such as an ocean. There are many interactions between living things within this ecosystem. Marine biologists study the population of the plants and animals in this ecosystem when they notice a change. Suppose a predator population suddenly decreased even though the prey population stayed the same. Besides disease, what could explain this change? Circle all that apply.

    A. The predator had its own predator whose population increased.

    B. The population of producers in the ecosystem died out.

    C. The population of prey stayed the same.

    D. A competing predator entered the ecosystem and is getting to the prey first, leaving the original predator without resources.

Extend It

Choose an ecosystem that appears near where you live. Identify an interaction between living things in the ecosystem. Write a poem about the interaction. Research more information about the interaction to add details to your poem.

_____

_____

_____

_____

_____

_____

_____

_____

_____

_____

_____

_____

_____

Memorize your poem, and recite it to a partner or to a small group. Your teacher may want you to recite your poem to the class.

KEEP PLANNING
STEM Module Project
Science Challenge

Now that you have learned about interactions of living things in ecosystems, go to your Module Project to use this information to help you design a compost heap.

Moldy Bread

Two friends were making lunch. They noticed the loaf of bread was moldy. They wondered what would happen to the moldy bread if they put it outside in the garden. This is what they said:

Miguel: *The matter that makes up the loaf of bread will eventually disappear and can't be used anymore.*

Shani: *The matter that makes up the loaf of bread will eventually change and be used in a different form.*

Who do you think has the best idea? _____

Explain your thinking.

_____

_____

_____

_____

_____

_____

_____

You will revisit the Page Keeley Science Probe later in the lesson.

Role of Decomposers

What is the relationship between the mushroom and the log?

⊙ GO ONLINE

Check out *Mushrooms* to see the phenomenon in action.

Look at the photo and watch the video of the mushrooms. What do you observe? What questions do you have about the phenomenon? Record or illustrate your thoughts below.

Did You Know?

Some mushrooms can absorb and digest dangerous substances like oil, pesticides, and industrial waste.

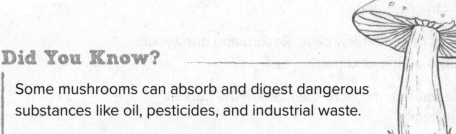

INQUIRY ACTIVITY

Hands On

Yeast and Bananas

Think about the mushroom on the log. You are going to investigate how decomposers affect matter in an ecosystem over time.

Make a Prediction How will yeast affect a banana over time?

_____

_____

_____

_____

_____

_____

_____

Materials

4 small pieces of banana

2 resealable plastic bags

marker

1 package of yeast

plastic knife

paper towel

pan balance

Carry Out an Investigation

BE CAREFUL Wear gloves. Wash your hands after handling the lab materials.

1. Place two banana pieces of the same size into two separate plastic bags.

2. Sprinkle the package of yeast on one of the pieces of banana. Then, close the bags. Label one "with yeast" and one "without yeast" and keep the bags closed.

3. Measure the mass of the banana pieces in the plastic bags. Record the data in the table on the next page.

4. Observe the bananas over the next few days. Record and draw your observations in the tables on the next page.

5. On the last day of observations, measure the mass of the banana pieces again. Record this data.

Record Data

Time (Days)	Banana with Yeast	Banana without Yeast
1		
2		
3		
4		
5		

	Beginning Mass	End Mass	Beginning Illustration	Ending Illustration
Banana without Yeast				
Banana with Yeast				

Communicate Information

Compare your results with a partner's results. Did your results support your prediction? How did the yeast affect the banana over time?

VOCABULARY

Look for these words as you read:

bacteria

decomposer

fungi

This fallen log is part of a tiny ecosystem that includes populations of fungi, moss, and bacteria.

The bacteria that live on the roots of plants change nitrogen gas into a form plants can use.

Role of Decomposers in Ecosystems

The plants and animals in an ecosystem have different roles based on how they feed. All ecosystems have organisms that are decomposers. Decomposition is the breaking down or decaying of plant and animal material.

Organisms that turn this material into simpler substances are called **decomposers**. They are organisms that break down plant and animal matter. Most bacteria and fungi are decomposers. **Bacteria** are a type of organism made up of a single cell.

Fungi appear in many forms. They include the yeast that makes holes in bread as it rises, the fuzzy mold that warns us not to eat old food, smelly mildew, and the mushrooms that pop up overnight on the forest floor. For many years, scientists thought fungi were plants because they didn't move and many sprouted from soil. Unlike plants, however, fungi cannot make their own food.

Decomposers break down plant and animal matter, which returns nutrients to the soil. Growing plants then use these nutrients to survive. Making nutrients available to plants is the most important role of decomposers in an ecosystem.

ENVIRONMENTAL ▶ Connection

1. What factors can interfere with the role of decomposers?

_____

_____

_____

_____

_____

_____

Decomposers

🡒 **GO ONLINE** Watch the video *Decomposers*. Talk about what you observe with a partner.

WRITING Connection

Write an explanation on how **decomposers help an environment stay healthy and balanced** by contributing to the **movement of matter in ecosystems**.

_____

_____

_____

_____

_____

_____

_____

_____

_____

_____

_____

Fungi come in many shapes and sizes.

INQUIRY ACTIVITY

Hands On

Soil Decomposers

Soil decomposers work to break down dead matter. Different environmental factors affect how decomposers work. You are going to investigate how moisture affects decomposers.

Make a Prediction Are decomposers more active in damp or dry environments?

_____

_____

_____

_____

Carry Out an Investigation

BE CAREFUL Wear gloves. Wash your hands after handling the lab materials.

1. Plan your investigation below. Remember to take accurate measurements and include a control group in your investigation.

_____

_____

_____

_____

_____

_____

_____

Materials

4 small pieces of carrots

4 resealable plastic bags

soil

graduated beaker with water

2. Record your observations of the different carrots over time.

Record Data

	30 mL water	15 mL of water	No water	No soil
Day 1				
Day 4				
Day 7				
Day 10				

 Talk About It

Compare your results with a partner's results. Did your results support your prediction? How can you improve your investigation?

 Revisit the Page Keeley Science Probe on page 37.

What Does a Mycologist Do?

Mycologist

Reminder: Never eat mushrooms that you find growing in the wild!

With a magnifying glass and a pocketknife, mushroom scientist Greg Mueller is going on a treasure hunt. "I never know what I might find," he says. "And what I find today may be different four days from now." Mueller is a mycologist. A mycologist is a scientist who studies fungi.

Mueller explains that fungi are often present even when we don't see them. Fungi grow in hair-like threads. They spread through soil, rotting wood, or wherever a fungus seeks water and food. A single strand is too small to see with the naked eye. Mueller points out white spots on a fallen tree where many strands have grown together, creating a web.

Fungi come in an amazing array of shapes, colors, and sizes. Many aren't mushroom-shaped at all, such as Chanterelle, Slime mold, and Bracket fungus.

All fungi play one of three roles: some are decomposers, some form partnerships with living plants, and some are harmful to living plants. Most fungi Mueller happens to see on his walks are breaking down dead plant matter. This recycles the plant nutrients back into the soil. As fungi spread through a fallen tree to gather food, they destroy the stiff cell walls of the wood. This makes the nutrients inside available.

Mueller breaks a chunk of decaying wood from a tree, and it almost crumbles to sawdust in his fingers. That's a sign that fungi have done their work. "We'd have piles of dead trees miles high if we didn't have fungi," he says.

Fungi fill the forest, but they're easiest to spot after it rains, when mushrooms pop up.

It's Your Turn

How could a mycologist work with a horticulturist to improve gardening practices?

Review

EXPLAIN
THE PHENOMENON

What is the relationship between the mushroom and the log?

Summarize It

Explain why decomposers are important to the environment.

_____

_____

_____

_____

_____

_____

_____

REVISIT
PAGE KEELEY
SCIENCE
PROBES

Revisit the Page Keeley Science Probe on page 37. Has your thinking changed? If so, explain how it has changed.

1. Decomposers are important to the health and balance of an ecosystem because they:

   A. prey on overpopulated animals

   B. break down plant and animal matter

   C. are food for plants

   D. produce oxygen for living things

ENVIRONMENTAL ▶**Connection**

2. How do decomposers help humans manage the amount of food waste they produce?

   A. Decomposers help food ripen.

   B. Decomposers cause food to get moldy.

   C. Decomposers help to break down food waste, like in compost heaps.

3. How does this affect how decomposers impact other parts of our environment?

   _____

   _____

   _____

   _____

   _____

   _____

Extend It

Think about the role of decomposers in a forest ecosystem. Think about ways that humans change forests. How does human activity interfere with the role of decomposers?

_____

_____

_____

_____

_____

_____

_____

_____

OPEN INQUIRY

What questions do you still have about the role of decomposers?

_____

_____

_____

Plan and carry out an investigation or research to find the answer to your question.

KEEP PLANNING
STEM Module Project
Science Challenge

Now that you have learned about the role of decomposers, go to your Module Project to consider this information as you design a compost heap.

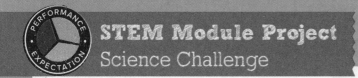
Design a Compost Heap

How can composting provide the nutrients that plants need to grow? You are going to think like a horticulturist and research different types of composting and related activities. Use your planning and research to develop a plan for a compost heap. You will write a proposal to implement a composting program at your school or home.

Planning after Lesson 1

Apply what you have learned about what plants need to live and grow to your project planning.

How does knowing about the needs of plants affect your project planning?

Planning after Lesson 2

Apply what you have learned about how living things interact to your project planning.

How would your project enhance the interactions among living things?

Planning after Lesson 3

Apply what you have learned about decomposers to your project planning.

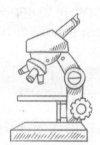

Describe the factors that you need to consider for decomposers to be successful at decomposing the material in the compost heap.

Use what you have learned to further research and plan the design for your project.

Define the Problem

What problems can you solve by implementing a compost heap in the garden?

Research the Problem

Research ideas for materials that you could use in your project by going online to teacher-approved websites or by finding books on composting in the school library.

Source	Information to Use in My Project

Sketch Your Model

Draw what you think the composting heap should look like. Use a separate piece of paper if needed. Include the length, width, and height of your compost heap.

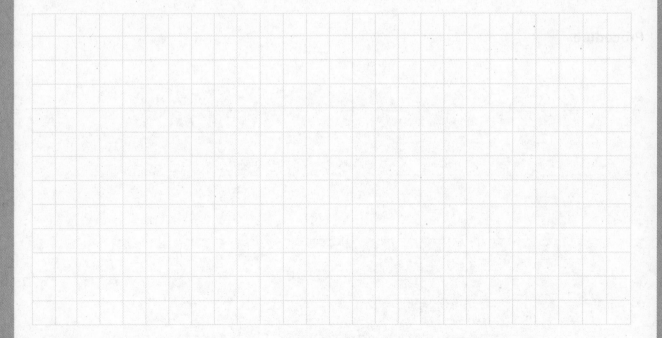

MATH Connection Calculate the volume of your compost heap. Show your work.

Design a Compost Heap

Look back at the planning you did after each lesson.
Use that information to complete your final module project.

Make a Plan

1. Use your project planning to make a plan for your compost heap.

2. Write clear steps for how your compost heap can be built.

3. Define the materials to build your compost heap. List the materials in the space provided.

4. Identify the number and type of abiotic and biotic components that will be included in your model.

Materials

_____

_____

_____

_____

_____

Procedure

_____

_____

_____

_____

_____

_____

_____

_____

Make Observations

Build your compost heap. Record your observations over time.
Use a data table if needed.

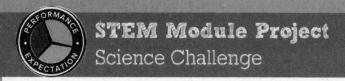

STEM Module Project
Science Challenge

Communicate Your Results

Think about what you learned about decomposers and interactions of living things in ecosystems. Write a proposal about why you think your school or home should compost food waste. Draw a diagram of the location of the compost heap.

_____

_____

_____

_____

_____

MATH Connection   On your diagram, indicate the distance between your compost heap and the nearest building in the area. What is the estimated distance in meters? How about in kilometers?

MODULE WRAP-UP

REVISIT
THE PHENOMENON

Using what you learned in this module, explain what happens to matter in ecosystems over time.

Have your ideas changed? Explain.

Energy in Ecosystems

ENCOUNTER
THE PHENOMENON

How is the whale getting and using energy?

🕭 GO ONLINE

Check out *Baleen Whales Feeding* to see the phenomenon in action.

💬 Talk About It

Look at the photo and watch the video of the baleen whale as it eats. What questions do you have about the phenomenon? Talk about them with a partner.

Did You Know?

During its feeding season, some baleen whales eat up to 4 tons, or 8,000 pounds, of food each day. A human eats the same amount of food over the course of 4 years!

Build an Eco-Column

How can we model the flow of energy in an ecosystem? You are an ecological engineer who wants to build an eco-column to use when students come for field trips to your lab. An eco-column is a model of an entire ecosystem on a smaller scale. At the end of this module, you will build your model to show how energy flows through all of Earth's systems.

Lesson 1
Earth's Major Systems

Lesson 2
Cycles of Matter in Ecosystems

Lesson 3
Energy Transfer in Ecosystems

Ecological engineers study ecology as well as engineering. Ecology is the study of how living things interact in their environments. Ecological engineers often design or restore ecosystems so that humans and other living things can both benefit.

STEM Module Project

Plan and Complete the Science Challenge Use what you learn to design a successful eco-column.

Earth's Systems

Four friends were talking about Earth's systems. They each had a different idea about what they are made up of. This is what they said:

Tobias: I think Earth's systems include the materials on Earth that interact.

Jason: I think Earth's systems include the materials on land and water only.

Fay: I think Earth's systems include the processes and materials on Earth that interact.

Kristy: I think Earth's systems include the processes and materials on Earth that interact as long as they are on land or in water.

Who do you agree with most? _____

Explain why you agree.

_____

_____

_____

You will revisit the Page Keeley Science Probe later in the lesson.

Earth's Major Systems

This U-shaped valley is located in Yosemite National Park.

What types of matter do you see in Yosemite National Park?

Yosemite National Park

◑ GO ONLINE

Check out *Yosemite National Park* to see the phenomenon in action.

Look at the photo and watch the video about the U-shaped valley in Yosemite National Park. What did you observe? What questions do you have about the phenomenon? Record your thoughts below.

_____

_____

_____

_____

_____

_____

_____

Did You Know?

U-shaped valleys are usually formed by the movement of giant sheets of ice called glaciers.

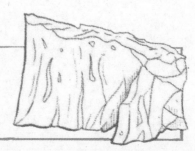

INQUIRY ACTIVITY

Materials

photo
cards

Data Analysis

Types of Matter on Earth

Think about the types of matter in Yosemite National Park. Use
the photos to think about the types of matter found on Earth.

Make a Prediction How can we categorize the different
types of matter on Earth?

_____

_____

_____

_____

Carry Out an Investigation

1. Use photos provided by your teacher. Categorize,
   or group, the photos based on what is similar.

2. Record how you organized the photos in the box below.

Communicate Information

3. Compare how you organized the photos with another group of students. How were the methods of organizing the photos the same? How were they different?

_____

_____

_____

_____

4. How could you sort the photos to fit into only four categories? Record how you would organize them below.

5. Do your results support your prediction? Explain.

_____

_____

_____

_____

💬 Talk About It

Do some of the photos fit into more than one category?
Talk about it with a partner.

Earth's Systems

VOCABULARY

Look for these words as you read:

atmosphere

biosphere

geosphere

hydrosphere

The parts that make up Earth can be organized into four main systems. Systems are a collection of different components that all work together.

The **atmosphere** is a layer of gases surrounding Earth. Made up mostly of nitrogen and oxygen, the atmosphere also contains water vapor, carbon dioxide, and other gases.

The **geosphere** includes the solid and molten rock inside Earth. It also includes the soil, rock pieces, and land features at Earth's surface. Hills, mountains, erupting volcanoes, and other landforms are all part of the geosphere.

All of Earth's liquid and solid water, including oceans, lakes, rivers, glaciers, and ice caps, makes up the **hydrosphere**. The hydrosphere covers more than 70 percent of Earth's surface. It exists in two forms: salt water and fresh water. Most of Earth's fresh water exists as ice. Most of Earth's salt water is in the ocean.

The **biosphere** is all of Earth's living things. Organisms that make up the biosphere are found from the lower atmosphere to the depths of the ocean floor. All living things are part of the biosphere.

GO ONLINE Watch the video *Four Earth Systems* to see some ways that these systems affect each other.

REVISIT Revisit the Page Keeley Science Probe on page 61.

PAGE KEELEY SCIENCE PROBES

Identifying Earth's Systems

FOLDABLES®

Cut out the Notebook Foldables given to you by your teacher.
Glue the anchor tabs as shown below. Use what you have learned to
describe each of the four systems and provide examples of each.

Glue the anchor tab here.

INQUIRY ACTIVITY

Earth's Systems
Within Ecosystems

State the Claim How can we make a model of Earth's four systems?

_____

_____

_____

 Think about **the four systems on Earth.** Choose an ecosystem.
Develop a model of the ecosystem using what you have learned
about Earth's systems. Draw your model in the space below.

Use Your Model

1. Switch models with a partner. Label the four systems on your partner's model. Trade your models back. Was your partner successful at labeling the four systems in the ecosystem you modeled? Explain.

_____

_____

_____

_____

2. **ENVIRONMENTAL** ⟩**Connection** Are all four systems important in an ecosystem? Why or why not?

_____

_____

_____

_____

_____

3. How can we define the different types of matter that are on Earth?

_____

_____

_____

_____

What Does a Systems Engineer Do?

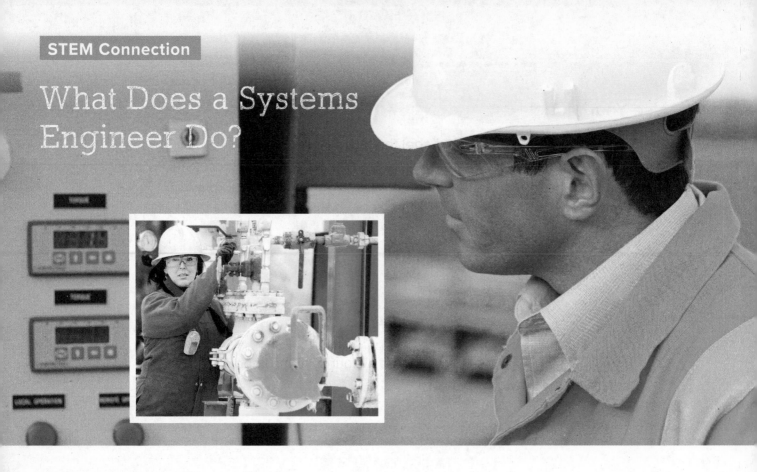

Systems are everywhere. They appear in nature, but they also appear in human-made items such as computers. Remember that systems are a collection of different components that all work together. Each of the parts need to work correctly to be successful. **Systems engineers** help to design and organize different types of systems that are needed to solve a problem. These engineers work with other types of engineers depending on the desired outcome. They could be helping to build computer software, distribute goods within a community, or working with complex machinery. They make sure that the system works correctly throughout the entire problem-solving process. If you like making things work, you might want to become a systems engineer.

It's Your Turn

How might an ecological engineer and a systems engineer work together? What advice might a systems engineer have about building an eco-column?

Word Origins

How can parts of words help us understand the whole meaning of a word?

_____

_____

1. Look at the lesson vocabulary words. What do you notice about the words?

_____

_____

2. Research the word origins of the lesson vocabulary words. Record information about each in the box below.

Word	Prefix Meaning	Suffix Meaning
atmosphere		
biosphere		
geosphere		
hydrosphere		

💬 Talk About It

How can understanding parts of scientific words help us understand other words?

Review

EXPLAIN
THE PHENOMENON

What types of matter do you see in Yosemite National Park?

Summarize It

Use what you have learned to communicate what you might find in each Earth system pictured above.

_____

_____

_____

_____

_____

_____

_____

REVISIT
PAGE KEELEY SCIENCE PROBES

Revisit the Page Keeley Science Probe on page 61. Has your thinking changed? If so, explain how it has changed.

 **Three-Dimensional Thinking**

1. Which of the following is *not* part of Earth's geosphere?

   A. mountains

   B. soil

   C. rivers

   D. volcanoes

2. The biosphere contains all of the nonliving things on Earth.

   A. True

   B. False

3. The _____ includes all the gases around the Earth.

   A. hydrosphere

   B. atmosphere

   C. geosphere

   D. biosphere

4. You are walking through Yosemite National Park. What would you see that are all part of the biosphere?

   A. Pacific tree frogs, mountain chickadees, cumulus clouds

   B. Sierra mountain kingsnake, sedimentary rocks, dragonflies

   C. black bears, freshwater ponds, trout

   D. dragonflies, Pacific tree frogs, black bears

Extend It

Think about what Earth's systems might look like in other parts of the world compared to where you live. What do you think is the same and different about Earth's systems in another country? Choose a specific example to compare to your region by researching information or using a world map.

_____

_____

_____

_____

_____

_____

_____

_____

_____

_____

_____

_____

_____

_____

KEEP PLANNING

STEM Module Project
Science Challenge

Now that you have learned about Earth's major systems, go to your Module Project to consider this information as you make your plan to build an eco-column.

LESSON 2 LAUNCH

What Happens to the Matter?

Two friends disagreed about what happens to the matter in an ecosystem as organisms die and living organisms eat, breathe, and drink water. This is what they said:

Baye: *I think the matter gets recycled in the living and nonliving things in an ecosystem.*

Arlen: *I think the matter eventually gets used up and then is gone from the ecosystem.*

Who do you agree with the most? _____

Explain why you agree.

_____

_____

_____

_____

_____

You will revisit the Page Keeley Science Probe later in the lesson.

Cycles of Matter in Ecosystems

ENCOUNTER
THE PHENOMENON

Where does the water in this hot spring come from?

GO ONLINE

Check out *Yellowstone Hot Springs* to see the phenomenon in action.

Look at the photo and watch the video of the hot springs at Yellowstone National Park. What questions do you have about the phenomenon? Record or illustrate your thoughts below.

Did You Know?

One type of hot spring is a geyser. Yellowstone is home to more than half of all of the geysers in the world!

INQUIRY ACTIVITY

Hands On

Cycling of Matter

Think about the hot spring and what you have learned about how matter changes state.

Make a Prediction How can water travel inside a bowl covered in plastic wrap?

_____

_____

_____

_____

_____

Materials

clear container with water

plastic wrap

large rubber band

ice cubes

Carry Out an Investigation

1. Fill the bowl ¼ full with water.

2. Place the plastic wrap over the top of the container. Use a rubber band to secure the plastic wrap.

3. Place several ice cubes in the center of the top of the plastic wrap. Place the container in a sunny area.

4. Check on the container every five minutes. Record your observations.

Record Data

Time (Minutes)	Observations
0	
5	
10	
15	

💬 Talk About It

Compare your results with a partner's results. Was your prediction supported by your results? Is there any evidence that matter was conserved in this system? What could you do to find out?

INQUIRY ACTIVITY

Communicate Information

MATH Connection Not all of California receives the same amount of rainfall each year. On a separate piece of paper, draw a bar graph to represent the data.

City	Annual Rainfall Averages from 1981–2010
Sacramento	470 mm (18.5 in)
San Francisco	601 mm (23.7 in)
Fresno	292 mm (11.5 in)
Los Angeles	379 mm (14.9 in)
San Bernardino	407 mm (16.0 in)
San Diego	263 mm (10.3 in)

 Use the data represented in the bar graph about average rainfall of six cities on the map. Explain any **patterns** you see in the data. Which **Earth's system** includes rainwater as one of its components?

_____

_____

_____

_____

_____

_____

_____

MAKE YOUR CLAIM

How does matter in an ecosystem cycle?

Make your claim. Use your investigation.

CLAIM

Matter cycles through an ecosystem _____.

Cite evidence from the activity.

EVIDENCE

The investigation showed _____.

Discuss your reasoning as a class. Tell about your discussion.

REASONING

The evidence supports the claim because _____.

You will revisit your claim to add more evidence later in this lesson.

Water Cycle

VOCABULARY

Look for these
words as you read:

condensation

evaporation

nitrogen cycle

oxygen-carbon
cycle

precipitation

runoff

water cycle

The **water cycle** is the continuous movement of water between Earth's surface and the air, changing forms among the three states of matter.

The Sun is the energy source for the water cycle. The Sun's energy causes water to evaporate. When water **evaporates,** it changes from a liquid to a gas in the form of water vapor. Water is constantly evaporating from the leaves of plants. This process is called transpiration. Transpiration is one way water vapor returns to the atmosphere.

As it rises in the air, the water vapor cools and condenses. **Condensation** occurs when a gas changes to a liquid. Water vapor can condense on dust particles in the air, forming clouds. Inside a cloud, small water droplets may combine and form larger ones. Some droplets freeze into ice if conditions allow it. During **precipitation,** water falls from clouds over land and water. Precipitation can fall as rain, sleet, snow, or hail.

When it rains, water flows over Earth's surface as **runoff**. Runoff gathers in lakes, oceans, and streams. Water that soaks into the ground moves downward through small cracks and spaces and is called groundwater. Plants take in groundwater from the soil through their roots.

> **GO ONLINE** Watch the video *Water Cycle* to learn more about precipitation, condensation, and evaporation.

Label evaporation, condensation, and precipitation in the diagram below.

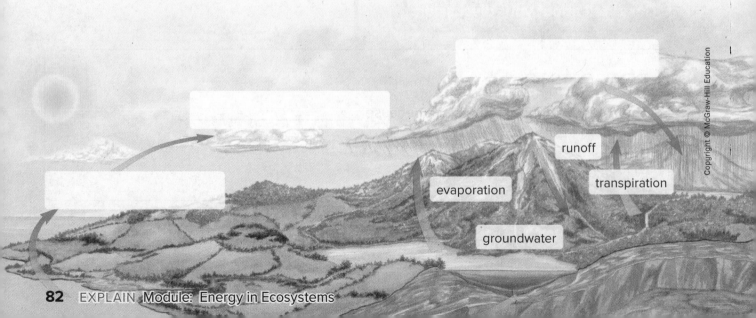

runoff

evaporation

transpiration

groundwater

Copyright © McGraw-Hill Education

Nitrogen Cycle

Air is made up of 78 percent nitrogen, but few living things can use nitrogen gas. First, nitrogen must be fixed, or changed into a form that living things can use. The **nitrogen cycle** is the continuous circulation of nitrogen from air to soil to organisms and back to air or soil.

Some bacteria that live on roots of plants can change nitrogen gas into a form plants can use. Nitrogen can come from fertilizers that are added to soil. Nitrogen can also be fixed by volcanic activity and lightning.

As plants grow, they absorb this form of nitrogen to make proteins. When animals eat plants or other plant-eating animals, they take in the stored nitrogen.

Nitrogen is eventually released into the soil through animal waste and decayed plants and animals. Decomposers and bacteria help return nitrogen into the atmosphere, and the cycle repeats.

How does nitrogen flow through an ecosystem? Draw a diagram of the process in the box below.

⊘ **GO ONLINE** Explore *Nitrogen in Ecosystems* to see the nitrogen cycle within an ecosystem.

Nodules can appear on roots where bacteria change nitrogen.

Oxygen-Carbon Cycle

Plants and animals are both part of the oxygen-carbon cycle.
The **oxygen-carbon cycle** is the circulation of oxygen and carbon
dioxide gas. Producers give off oxygen, as well as some carbon dioxide.
When consumers use energy, they release carbon dioxide as waste.

Millions of years ago, Earth's atmosphere had no breathable oxygen.
Early producers—such as plants—used water, carbon dioxide, and energy
from the Sun to make their own food. Over time, the oxygen released by
producers gradually built up in the atmosphere. Today, producers
continue to cycle oxygen into the atmosphere. Carbon dioxide is a
waste product produced by all living things, including plants, when
they use energy.

Activities such
as the burning of fossil
fuels also release
carbon dioxide.

As producers,
plants take in carbon
dioxide and give off
oxygen. Animals take in
oxygen and give off
carbon dioxide.

GO ONLINE Watch the video
Terrariums to learn more about how
matter moves through an ecosystem.

COLLECT EVIDENCE

Add evidence to your claim on page 81 about how matter
cycles within an ecosystem.

Read the Investigator article *Super Corals* and other resources about an ecosystem where the three main cycles are no longer balanced.

WRITING Connection Write a report about the effects on the ecosystem and what is being done to help.

_____

_____

_____

_____

_____

_____

_____

_____

_____

_____

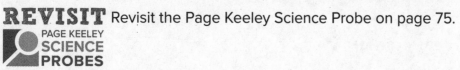

REVISIT Revisit the Page Keeley Science Probe on page 75.

PAGE KEELEY SCIENCE PROBES

FOLDABLES®

Cut out the Notebook Foldables tabs given to you by your teacher. Glue the anchor tabs as shown below. Describe how the three cycles you have learned about work in a rain forest ecosystem.

Glue the anchor tab here.

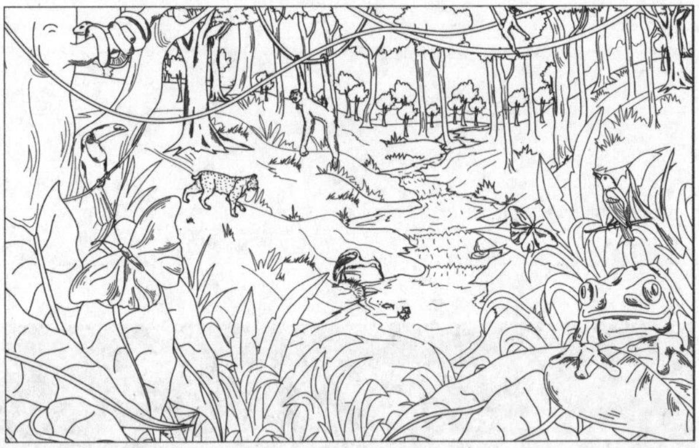

A Day in the Life of a Landfill Manager

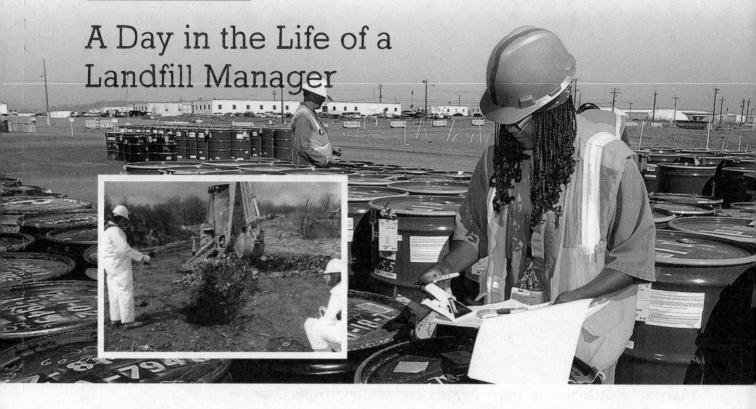

Landfill managers help cycle matter that humans throw away so the matter does not harm ecosystems. They have the important job of maintaining the materials that are placed in a landfill, since everything cannot be recycled. Most landfill managers have certification through the Solid Waste Association of North America. That certification allows them to make sure the landfill can meet the regulations of the Environmental Protection Agency. This confirms that the landfill is not polluting or causing harm to nearby waterways or habitats. Landfill managers operate complex machinery as well as train other people. It can be a messy job, but it's an important one!

It's Your Turn

How can a landfill manager use information about cycles in ecosystems to make sure they are able to protect nearby land and water?

Review

EXPLAIN
THE PHENOMENON

| Where does the water in this hot spring come from?

Summarize It

Explain how matter such as water, oxygen, and carbon dioxide cycles in an ecosystem.

_____

_____

_____

_____

_____

_____

_____

_____

REVISIT
**PAGE KEELEY
SCIENCE
PROBES**

Return to the Page Keeley Science Probe on page 75. Has your thinking changed? If so, explain how it has changed.

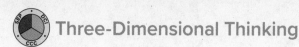 **Three-Dimensional Thinking**

1. The water cycle includes water evaporating into water vapor which can form clouds.

　A. True

　B. False

2. Construct an explanation about the role that bacteria play in the nitrogen cycle.

_____

_____

_____

_____

_____

_____

3. **ENVIRONMENTAL Connection** How does conserving the amount of water we use affect the water cycle? Circle all that apply.

　A. limits the amount of water we remove from natural waterways

　B. allows natural water levels to remain at healthy levels

　C. adds pollution to nearby waterways

Extend It

ENVIRONMENTAL **Connection** During a drought, less water than normal falls as precipitation. How does this affect the economy in an area that is experiencing drought? Research information to help you respond.

_____

_____

_____

_____

_____

_____

_____

_____

_____

_____

_____

_____

_____

_____

_____

KEEP PLANNING

STEM Module Project
Science Challenge

Now that you have learned about the cycles of matter in ecosystems, go to your Module Project to consider this information as you build an eco-column.

Energy and Matter in Ecosystems

Three friends were talking about how matter and energy move through living things in an ecosystem. They each had a different idea. This is what they said:

Ted: *I think matter gets recycled in ecosystems, but energy does not get recycled.*

Jill: *I think energy gets recycled in ecosystems, but matter does not get recycled.*

Lee: *I think both matter and energy get recycled in ecosystems.*

Whom do you agree with most? _____

Explain why you agree.

_____

_____

_____

_____

You will revisit the Page Keeley Science Probe later in the lesson.

Energy Transfer in Ecosystems

How do these birds get energy from their ecosystem?

GO ONLINE

Check out *Sanderlings on the Beach* to see the phenomenon in action.

Look at the photo and watch the video of the sanderlings hunting for food. What questions do you have about the phenomenon? Record or illustrate your thoughts below.

Did You Know?

Sanderlings breed in the Arctic region of the world and migrate as far south as Australia. When they need to escape a predator in flight, they may dive into the water.

INQUIRY ACTIVITY

Hands On

Ecosystem Tag

Think about the bird eating on the beach. All living things need to get energy from their ecosystem to survive. You have learned about the relationship of predators and prey in an ecosystem.

Make a Prediction How is energy transferred from prey to predator?

_____

_____

_____

_____

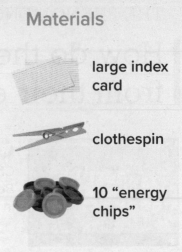

Materials

large index card

clothespin

10 "energy chips"

Carry Out an Investigation

BE CAREFUL Tap your classmates on the shoulder to tag them.

1. Your teacher will give you an index card with the name of an organism on it. Attach the card to your shirt using the clothespin.

2. You will also receive ten "energy chips." This represents the amount of energy you have.

3. Follow your teacher's instructions. One end of the game area will be shelter for krill. Krill cannot stay in the shelter for more than 10 seconds.

4. When your teacher tells you to start, try to tag your prey. Collect one energy chip from each prey you tag.

   • Whales can tag salmon, herring, or krill.

   • Salmon can tag herring or krill.

   • Herring can tag krill.

   • Krill need to avoid predators and get to shelter to collect one more energy chip.

5. When a predator tags you, give them one of your energy chips. Don't tag a prey that is already stopped to give an energy chip to another predator.

6. Once you run out of energy chips, you are out of the game.

Communicate Information

7. Did your results support your prediction? Explain.

_____

_____

_____

_____

_____

_____

_____

_____

_____

_____

_____

_____

_____

_____

💬 Talk About It

Talk about what you saw in the activity with a partner. Choose a group of animals from another ecosystem that could model these same types of interactions.

Look for these
words as you read:

consumer

energy flow

food chain

food web

producer

Organisms in Ecosystems

Producers Every organism in an ecosystem relies on producers. A **producer** is an organism that uses energy from the Sun to make its own food. Producers include green land plants such as grasses and trees. Algae are the main producers in lakes and oceans.

Indian clover are a type of plant, which means they are producers and make their own food.

Consumers Organisms that cannot make their own food are called **consumers**. Animals are consumers. Recall that consumers are classified by the kind of food they eat. Herbivores eat only plants. Carnivores eat only other animals. Omnivores are animals that eat both plants and animals for energy.

These quail are consumers. They eat mostly seeds, but will eat plants and small insects.

Decomposers Organisms that break down dead or decaying plant and animal materials are decomposers. You have learned how important decomposers are to the complex cycles in ecosystems.

GO ONLINE Explore *Energy Transfer in a Food Chain* to observe how organisms obtain and use energy in their ecosystem.

Think about the animals from the *Ecosystem Tag* activity. How can you show the direction of energy transfer among the organisms from the activity? Draw your ideas in the box below.

Inspect

Read the passage *Food Chains and Food Webs.* Underline the text evidence that tells where all energy in a food chain begins.

Find Evidence

Reread Where does the mayfly get energy? Highlight the energy source or sources.

Notes

_____

_____

_____

_____

_____

_____

_____

_____

_____

_____

_____

_____

Food Chains and Food Webs

The path that energy and nutrients follow is called a **food chain**. Food chains model a series of feeding relationships among organisms in an ecosystem. The **energy flow** in a food chain only goes in one direction because the source of all energy in a food chain is the Sun. This energy is used by producers, which are the base of all food chains. From there, energy moves from producers to consumers. All food chains end with decomposers.

Pond Food Chain

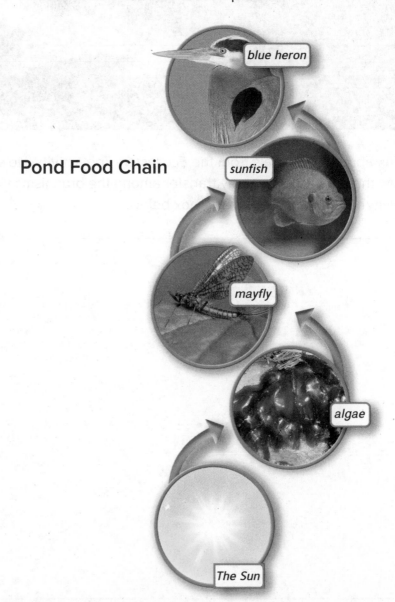

blue heron

sunfish

mayfly

algae

The Sun

A **food web** is a model that shows how the food chains in an ecosystem are linked together. Just as with food chains, arrows show the direction of the energy flow from one organism to another. The food web shows the predators and prey in an ecosystem. A predator may feed upon more than one type of prey for energy, and a prey may have more than one type of predator. This keeps the ecosystem balanced. In a balanced ecosystem, all living things can get energy they need to survive. This includes energy for growth, motion, body repair, and warmth.

Land Food Web

1. **Read a Diagram** How are the snake and mountain lion connected in the land food web?

_____

_____

_____

_____

_____

Make Connections

💬 Talk About It
Make connections to what you have learned about the role of decomposers. How do producers and consumers need decomposers to get the energy they need to survive?

Notes

_____

_____

_____

_____

_____

_____

_____

_____

_____

_____

_____

INQUIRY ACTIVITY

Research

Model a Food Chain and a Food Web

Choose and write down an ecosystem to research. You will make a model of a food chain and a food web within that ecosystem.

_____

State the Claim How does energy flow in the ecosystem you will research?

_____

_____

_____

_____

1. Use print and online resources to obtain information about the living things in the ecosystem you want to research. Record the information below. Include information about where each living thing gets its energy.

Ecosystem:		
Plants	**Animals**	**Decomposers**

Copyright © McGraw-Hill Education

2. Use the information you collected to plan your model of two food chains in the ecosystem. In the space below, draw a sketch of the flow of matter and energy between the living things. Remember to show the source of energy in the ecosystem.

 Now, on a separate sheet of paper, **develop a model** of a food web of the two food chains you drew. Show **connections among the living things** that represent the **flow of matter and energy**.

💬 Talk About It

Did your research and the models you developed support your claim? Talk about it with a partner.

> ⓥ **GO ONLINE** Explore *Energy Flow* to confirm your understanding of how energy flows in a food chain.

  Revisit the Page Keeley Science Probe on page 91.

What Does a Wildlife Conservationist Do?

When **wildlife conservationist** and tiger researcher Ullas Karanth was a young boy living in a village in southwest India, he saw wildlife everywhere. But as Ullas grew up, he saw forests cut down all around India. Because animals cannot survive when their habitats are destroyed, Ullas worried that the animals he loved would soon disappear. Indeed, tigers were already vanishing. In all his years of looking, Ullas had never seen one in the wild.

In India, some people share the forests with tigers, living in small villages, collecting wood for fires, and hunting deer and other wild animals. They compete with tigers for food. Areas that have been overhunted are called empty forests. Without food to eat, tigers in empty forests starve. People also compete with tigers for land. Farmers cut down forests and plow up grasslands to make bigger farms. Governments clear land to build roads and dams. The result is that only a tiny bit of land is left for the tigers. "Freeing up space for tigers is the single biggest challenge we have," says Karanth. "We can't have people grazing their livestock or going into the forests to kill deer and pig if we want to have tigers."

In the past, most wild animals existed without any human help. But with so few tigers left today, people have to step in and protect the animals to prevent them from disappearing forever. Karanth believes that science provides the best tools for saving these endangered cats. He hopes that studying tigers can give scientists and conservationists the information they need to work with governments to save the tigers and their forests.

Ullas Karanth believes that we should protect tigers because every species on the planet is connected. This means that in some way, we too depend on tigers, as we do on all species in the web of life. He explains that if people were to destroy the famous Indian building called the Taj Mahal, it could be rebuilt. "But once we destroy all these intricate ecological webs," he says, "there's no bringing them back!"

ENVIRONMENTAL ▶Connection

Why is deciding to help conserve wildlife an important yet difficult decision to act on? Use examples from what you read to help you respond.

_____

_____

_____

_____

_____

_____

_____

_____

_____

▶ **GO ONLINE** Use the Personal Tutor *Food Web* to learn more about the connections within food webs.

Review

EXPLAIN
THE PHENOMENON

How do these birds get energy from their ecosystem?

Summarize It

Explain how energy in the food an animal eats was once energy from the Sun. Describe how the energy is used by living things in the ecosystem.

_____

_____

_____

_____

_____

_____

_____

_____

REVISIT

PAGE KEELEY SCIENCE PROBES

Revisit the Page Keeley Science Probe on page 91. Has your thinking changed? If so, explain how it has changed.

Three-Dimensional Thinking

1. Describe how energy is transferred as it flows through an ecosystem.

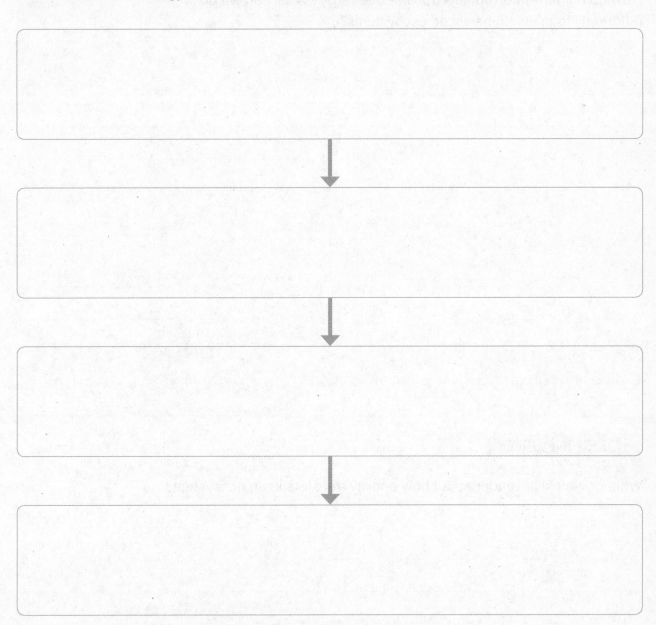

2. Why are producers the first organisms in a food chain?

    A. They prey on all other organisms.

    B. They receive energy directly from the Sun.

    C. They are not consumed by other organisms.

    D. They break down dead plant and animal matter.

Extend It

Food chains and food webs model the transfer of energy in ecosystems. How do humans interrupt this transfer of energy? What can we do differently to maintain balanced ecosystems?

_____

_____

_____

_____

_____

_____

_____

_____

_____

OPEN INQUIRY

What do you still wonder about how energy transfers in an ecosystem?

_____

_____

_____

Plan and carry out an investigation or research to find the answer to your question.

KEEP PLANNING
STEM Module Project
Science Challenge

Now that you have learned about the energy transfer in ecosystems, go to your Module Project to consider this information as you build an eco-column.

Build an Eco-Column

How can we model the flow of energy in an ecosystem? You are an ecological engineer who wants to build an eco-column to use when students come for field trips to your lab. An eco-column is a model of an entire ecosystem on a smaller scale. Use what you have learned throughout the module to plan an eco-column that will show how energy flows through all of Earth's systems.

Planning after Lesson 1

Apply what you have learned about Earth's major systems to your project planning.

Think about the parts of your eco-column. What parts appear in each of Earth's systems?

Atmosphere	Biosphere	Geosphere	Hydrosphere

Planning after Lesson 2

Apply what you have learned about how matter cycles through ecosystems to your project planning.

How will the eco-column show cycles of matter within an ecosystem?

Planning after Lesson 3

Apply what you have learned about how energy is transferred in ecosystems to your project planning.

Predict how the eco-column will show the transfer of energy within the system.

Sketch Your Model

Use what you have learned throughout the module to sketch your plan for your eco-column. Label the components of your model and what you expect each of them to do once the model is built.

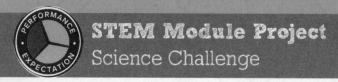
Build an Eco-Column

Look back at the planning you did after each lesson.
Use that information to complete your final module project.

Build Your Model

1. Make a plan to assemble your eco-column using your project planning pages.

2. Determine the materials you will need for your eco-column. List them in the space provided.

3. Follow your teacher's instructions to assemble your eco-column.

4. Each day, observe your eco-column. Look for the ways the hydrosphere, biosphere, geosphere, and atmosphere are present and interacting. Record your observations in the next section.

Materials

_____

_____

_____

_____

_____

_____

_____

_____

_____

_____

_____

_____

Make Observations

Observe your eco-column over several days. Record your observations in the box below. Use a data table if you need to.

STEM Module Project
Science Challenge

Communicate Your Results

As an ecological engineer, how could you use the eco-column that you built to teach students about the cycles in ecosystems?

MODULE WRAP-UP

REVISIT
THE PHENOMENON

Describe how the results of your project can help you explain the movement of energy and matter within an ecosystem, and how the whale is getting and using energy.

How have your ideas changed throughout the module?

_____

_____

_____

_____

Science Glossary

A

abiotic factor a nonliving part of an ecosystem

acid rain harmful rain caused by the burning of fossil fuels

air mass a large region of air that has a similar temperature and humidity

algae bloom a sometimes harmful rapid increase in the amount of algae found in water

apparent motion when a star or other object in the sky seems to move even though it is Earth that is moving

atmosphere the gases that surround Earth

B

bacteria a type of single cell organism

biosphere the part of Earth in which living things exist and interact

biotic factor a living thing in an ecosystem, such as a plant, an animal, or a bacterium

C

chemical change a change that produces new matter with different properties from the original matter

chemical property a characteristic that can only be observed when the type of matter changes

climate the average weather pattern of a region over time

colloid a type of mixture in which the particles of one material are scattered through another without settling out

condensation the process through which a gas changes into a liquid

conductivity ability for energy, such as electricity and heat, to move through a material

conservation the act of saving, protecting, or using resources wisely

conservation of mass a physical law that states that matter is neither created nor destroyed during a physical or chemical change

constellation any of the patterns of stars that can be seen in the night sky from here on Earth

consumer an organism that cannot make its own food

D

decomposer an organism that breaks down dead plant and animal material

deforestation the removal of trees from a large area

deposition the dropping off of eroded soil and bits of rock

E

endangered when a species is in danger of becoming extinct

energy the ability to do work or change something

energy flow the movement of energy from one organism to another in a food chain or food web

erosion the process of weathered rock moving from one place to another

evaporation a process through which a liquid changes into a gas

extinct when a species has died out completely

F

floodplain land near a body of water that is likely to flood

food chain the path that energy and nutrients follow among living things in an ecosystem

food web the overlapping food chains in an ecosystem

fungi plant-like organisms that get energy from other organisms which may be living or dead

G

galaxy a collection of billions of stars, dust and gas that is held together by gravity

gas a state of matter that does not have its own shape or definite volume

geosphere the layers of solid and molten rock, dirt, and soil on Earth

glacier a large sheet of ice that moves slowly across the land

gravity the force of attraction between any two objects due to their mass

groundwater water stored in the cracks and spaces between particles of soil and underground rocks

H

habitat a place where plants and animals live

hot spot an area where molten rock from within the mantle rises close to Earth's surface

hydrosphere Earth's water, whether found on land or in oceans, including the freshwater found underground and in glaciers, lakes, and rivers

I

ice caps a covering of ice over a large area such as in the polar regions

invasive species an organism that is introduced to a new ecosystem and causes harm

L

landslide the sudden movement of rocks and soil down a slope

light year the distance light travels in a year

liquids a state of matter that has a definite volume but no definite shape

M

magnetism the ability of a material to be attracted to a magnet without needing to be a magnet themselves

mass the amount of material in an object

matter anything that has mass and takes up space

meteor a chunk of rock from space that travels through Earth's atmosphere

meteorite A meteor that strikes Earth's surface

minerals solid, nonliving substances found in nature

mixture a physical combination of two or more substances that are blended together without forming new substances

molten rock very hot melted rock found in Earth's mantle

moon phases the apparent shapes of the Moon in the sky

N

nitrogen cycle the continuous circulation of nitrogen from air to soil to organisms and back to air or soil

O

orbit the path an object takes as it travels around another object

oxygen-carbon cycle the continuous exchange of carbon dioxide and oxygen among living things

P

phloem the tissue through which food from the leaves moves throughout the rest of a plant

physical change a change of matter in size, shape, or state that does not change the type of matter

physical property a characteristic of matter that can be observed and or measured

planet a large, round object in space that orbits a star

precipitation water that falls from clouds to the ground in the form of rain, sleet, hail, or snow

predator an animal that hunts other animals for food

prey animals that are eaten by other animals

producer an organism that uses energy from the Sun to make its own food

R

reflectivity the way light bounces off an object

reservoir an artificial lake built for storage of water

revolution one complete trip around an object in a circular or nearly circular path

rotation a complete spin on an axis

runoff excess water that flows over Earth's surface from a storm or flood

S

solid a state of matter that has a definite shape and volume

solubility the maximum amount of a substance that can be dissolved by another substance

solution a mixture of substances that are blended so completely that the mixture looks the same everywhere

star an object in space that produces its own energy, including heat and light

stomata pores in the bottom of leaves that open and close to let in air or give off water vapor

storage the process of water being stored on Earth's surface in the ground or as a water feature

T

tides the regular rise and fall of the water level along a shoreline

transpiration the release of water vapor through the stomata of a plant

V

volcano an opening in Earth's surface where melted rock or gases are forced out

volume a measure of how much space an object takes up

W

water cycle the continuous movement of water between Earth's surface and the air, changing from liquid into gas into liquid

weather the condition of the atmosphere at a given place and time

X

xylem the plant tissue through which water and minerals move up from the roots

Index

A

Abiotic factors, **26**
Agricultural and food
   science technicians, **14**
Atmosphere, **66**

B

Bacteria
   as decomposers, **42**
   defined, **42**
Bananas, and yeast, **40–41**
Biosphere, **66**
Biotic factors, **26**

C

Condensation, **82**
Consumers, organisms, **96**

D

Decomposers, **43**
   bacteria as, **42**
   defined, **42**
   fungi as, **42**
   organisms, **97**
   role in ecosystems, **42**
   soil, **44–45**

E

Earth's systems, **61**
   atmosphere, **66**
   biosphere, **66**
   geosphere, **66**
   hydrosphere, **66**

Eco-column, **107**
Ecological engineers, **60**
Ecology, **60**
Ecosystems
   abiotic factors, **26**
   biotic factors, **26**
   decomposers role in, **42**
   defined, **26**
   food chain, **98**
   food web, **99**
   habitats, **27**
   organisms in, **96–97**
   predators, **30**
   prey, **30**
   tag, **94–95**
Energy
   defined, **10**
   plants and, **10**
Energy flows, **98**
Engineers
   ecological, **60**
   systems, **70**
Entomologists, **32–33**
Evaporation, **82**

F

Food chain, **98**, **100–101**
Food web, **99**, **100–101**
Foxes, and rabbits, **22–24**
Fungi
   as decomposers, **42**
   defined, **42**

G

Geosphere, **66**

H

Habitats, **27**
Horticulturists, **4**
Hot spring, **88**
Hydrosphere, **66**

I

Invasive species, **28–29**
   defined, **28**
   introduction to
      environment, **28**
   spread of, **29**

L

Landfill managers, **87**

M

Matter
   cycling of, **78–79**, **81**
   nitrogen cycle, **83**
   oxygen-carbon cycle, **84**
   water cycle, **82**
Mueller, Greg, **46–47**
Mycologists, **46–47**

N

Nitrogen cycle, **83**

O

Organisms
   consumers, **96**
   decomposers, **97**

Dinah Zike's
Visual
Kinesthetic
Vocabulary®

✂ cut on all dashed lines

fold on all solid lines

biotic factors

food chain

A _____ is the overlapping food chains in an ecosystem.

A _____ is the path that energy and nutrients follow in an ecosystem.

Dinah Zike's
VKV
Visual
Kinesthetic
Vocabulary®

✂ cut on all dashed lines   ▢ fold on all solid lines

_____ are the effects on the ecosystem that are a result of the _____

nonliving parts of that ecosystem.

_____ are living things in an ecosystem, such as plants, animals, or bacteria.

Memory Maker: The prefix **a-** is a word part that sometimes means "not." If **biotic** means "related to living organisms," then what does **abiotic** mean? _____

a

web

Memory Maker: In your own words, explain the difference between a **food chain** and a **food web**.

water cycle

The _____ is the continuous trapping of nitrogen gas into compounds in the soil and its return to the air.

The _____ is the continuous exchange of carbon dioxide and oxygen among living things.

The _____ is the continuous movement of water between Earth's surface and the air, changing from liquid into gas into liquid.

Module: Energy in Ecosystems   **VKV3**

Dinah Zike's
VKV
Visual
Kinesthetic
Vocabulary®

✂ cut on all dashed lines   ⬒ fold on all solid lines

Memory Maker: Cycles are often represented with circles because, like circles, cycles are continuous. For each of the terms on this card, draw a circle diagram to define the term.

oxygen-
carbon dioxide

nitrogen